In memory of Chris Gammon: As beautiful as he was rare, A true friend to all…

Published by Lulu

ISBN: 978-1-4452-9357-8

Paradigm Shift:

The Future of Science & Technology

Paradigm Shift: Science & Technology of the Future

Try to not be drawn into this books small size and therefore fooled into believing that the stuff of scientific theory has to be overcomplicated with confusing equations and a verbose vocabulary that only the gifted or extremely well educated can understand; what is the point of creating a beautiful new theory that changes the paradigm (current interconnected theory, belief and experiment constituting the culture of science) of the human race if its understandings and wisdoms cannot be understood by everyone, surely a part of the degree of the truth of an idea or thought has is its relevance to the world as we understand it and hence in its understanding and practical value comes not only from the theories as the collection of theories, facts and truth but also proves its worth as a new paradigm of understanding through being understandable by the public and in its principles being accessible to anyone who has enough patience and determination to analyse, interpret, synthesise and deduce the fundamental forces and factors which determine the essential nature and intrinsic properties of the universe we inhabit.

It is my personal hope that some piece of theory or some idea that I have may be worthy of some award; be it either a PhD or a Nobel Prize, Well here is to my ambitions and desires (or

Paradigm Shift: Science & Technology of the Future

delusions) of greatness. As such when it comes to physics I am entirely self-trained as in school I failed to achieve a decent grade (achieving a grade D in Physics) due to the excessive oppression I suffered at the hands of many bullies who at one point made my life so unbearable that I was from my disposition of torment and despair contemplating suicide as a fourteen year old boy. This though had in the long term a positive role in my 'self development' as instead of suffering the abuse and the objects thrown at me from the back of the room I was often finding myself having enough of this bullying and was soon excusing myself from lessons somewhere in the first fifteen minutes with a similar feeling the teacher had for me as the extreme form of the bullying taking place in its intensity and complex dynamic leaving the teachers both helpless and powerless to do anything. On the positive side of things the intense emotions I was feeling found their form in the words and meanings of the poetry which I began writing as a response to the bullying where i found myself calm and alone in the cloakrooms of the school.

With the lack of a decent qualification in physics from my school days I was unable to pursue the subject further academically and instead focused my drive and ambition in the realm of philosophy which I studied academically for several years each year refining my understanding to an un-before reached plateau of knowledge and wisdom covering morality, the culture of science, the works of Karl Marx and John Stewart Mill and the philosophy of the fundamental nature of the world we inhabit in the many ways we inhabit it. Whilst developing a quicker intellect and more keen instincts for debate in the nature of the acquisition and causality of truth I came to make Philosophy with my reasonable success in the field the new Physics of my intellectual constitution but despite this I was still having deep conversations and meaningful debates with those who were studying physics picking up on interesting causalities and dynamics which exist in the essential structure and fundamental nature of the manifestation of physical reality.

After some success in my A-Level results I managed to secure a place at one of the more prestigious and elite of the universities of the British Isles at Cardiff and went in with the personal ideology that I was not there necessarily to get a degree; I figured that anybody with a bit of intelligence and some hard work if they pushed themselves could attain a degree of some class or other. As such my reason for a university education studying ideas so advanced that even today I am re-equating theories and theoretical understandings

Paradigm Shift: Science & Technology of the Future

from years ago into new ideological structures and paradigms which are the product of many years of intellectual refinement was in the rationale of my studies and my education an exercise in the acquisition of the essential facts and theories from the core body of Philosophy which could be used as individual particulars which link up together through causality to constitute a new inspired way of the future but born in its necessity in the modern day which through a conveyed understanding of its theory constitutes either a brand new paradigm or a new branch of scientific understanding which I hope revolutionises the nature and dynamics of the Modern world so that our children and our children's children and future generations have a legacy and a creed to inherit that would make us today proud instead of inheriting a world hell bent on weapons, oppression, fear, inequality, exploitation and untold misery.

After my formal education (which lead to the publishing of Dawn of the Neo-Modern: Art, Humanism & The Meme) at colleges and universities in the U.K. due to the overbearing stress and pressures of the university study coupled with the increased inner tension of some 'personal issues' that were a part of the mix of myself lead me down the path to the perfection of my manuscript and

4

negligence of my degree of chosen study through the profoundly negative inspiration that our life and our time may be up at any moment and judging the overbearing forces which were seeming to gather like storm clouds around me. As such all these energies and essences came to a head in me suffering a mental breakdown whilst being a student and it was in this time that I took to reading physics books (as it instilled within me a deep sense of calm, logical rationality) and started writing love poetry (as I figured that I would feel better after writing one hundred love poems then I was feeling before I had written them)

As I went on to read more and more physics books I found myself piecing together important theories, experiments and explanations gradually increased my theoretical ability and added confidence until one day I decided that I would have a go at explaining and getting theoretical with some of the more speculative and unsecured areas of Physics doctrine. Little did I believe that this intellectual pursuit would lead me to a grand overhaul of the great mass of Physics theory and the creation of what is if not a new paradigm of physics with its tenets, patterns and causal structures. Also if all my theoretical dabbling has not earned me the title of a 'cutting edge' radical scientist trail-blazing

Paradigm Shift: Science & Technology of the Future

a new ideological path to follow and a fresh new paradigm to explore (and hopefully be developed by others) then if somehow everything I have theorised, speculated and developed be either dismissible as subjective belief devoid of facts or complex conjecture which misrepresents current theories and understandings. At least in the initial designs in note books to the night I sat down in front of my computer and started the typing up of the soon published manuscript I had a lot of fun coming up with my explanations of different unaccounted for causalities and causal dynamics and theoretical deductions, inductions and generalisations and the associated facts and observations which were in there current state in my mind left more to be desired if a new formulation of the nature of the universe in a true unifying totalitarian theory is to ever come to exist in the essential formulation of the fundamental particulars of the grand explanation of Physics per se: the fundamental nature and dynamic of the universe we exist within as a part of. Here we go...

It was once coined and sometimes remarked that it is easier for a self-trained beginner at theoretical physics to have more of a chance of making successful developments and revisions to the subject then someone who is formerly trained in the subject. This theoretical person is as such someone who has studied Physics at an advanced or a degree level because the laws and dynamics, the theorems that define physics do have a current paradigm already; one with few anomalous results and unexplained causalities and such from this position impose a rigid structure over the observations and dynamics of the subject which don't get me wrong do work as explanations of the fundamentals of Physics have their place in the current paradigm and as such the theorems and debates as such have their place in the scientific culture of physics. As such the already developed relevant theory which as it exists in the modern day is such that it has needed some new postulations and theoretical designing as a part of the scientific culture of physics for some while. The current paradigm exists both as a cannon of scientific knowledge but also exists as a progressive self-developing theoretical design and as such the constituents of the current paradigm of physics in their resilience and persistence as robust parts of the scientists theoretical explanations and understandings cannot be denied. What here is the point I wish to make is that to the beginner the equating of physics theory as a painting which represents all theory and causality as brush-strokes when the explanations of this painting are removed from the contours of the artform then new and previously unknown, unperceived and unimagined causalities and causal patterns become more obvious and apparent to those of fresh minds; th. This freedom of interpretation and hence forth theorising and explanation by the self taught amateur physicist through their individual freedom from the contours, equations and theoretical dynamics of physics doctrine can discover new theory and experiment where the trained physicist would be too restrained and fixed in place to see. Through the unveiling of new theoretical dynamics or the development of paradigmatic weakness which causes the old theory to fail we move from one scientific epoch into another. As such this development of theory is all causal to the subsequent further development and evolution of experiment which is all based upon the current paradigm of scientific theory developing down new avenues and paths and in new directions which before were not even considered possible, that is the evolution of science. As such it is with new eyes and fresh minds that the novice can emerge discerning order from chaos in the tapestry of scientific culture and theory which can really be profound.

In order for my theory on the essential dynamics and causal patterns at work in the existence of a hyper-spatial eather to be truly understood and not confused in any way I shall first begin with the exposition of the theoretical dynamics of superstring theory. I shall then develop my arguments of the fundamental nature of reality through incorporating some of what is speculated in the field of zero-point energy which shall cement together the new paradigm and its theory. It is from these three different theoretical standpoints: from hyperspatial mechanics, superstring theory and zero-point energy with each having their own individual essential intellectual value and hence theoretical dynamics and developments. Through analysis and synthesis of these three areas which I will outline form and develop the new paradigm of scientific culture as each different area as I have set out with creates as a whole a synthesis creating a theory of hyper-spatially causal eather so fundamental to the way we see the physics and the world around us that it re-ignites the classical problems and dynamics of the subject of metaphysics which dates back to its creation in Ancient Greece by Aristotle.

Superstring theory is really an explanation of two things; it is an explanation of the dimensional essences and manifest forces involved in the creation of the universe at the moment of the big bang. Superstring theory is also an explanation of the fundamental nature of an atoms subatomic constituents: subatomic particles. As such regarding the size, electro-magnetic properties and the fundamental qualities, dynamics and causalities of these manifested and revealed sub-atomic particles with their own individual causal natures is such that they are perfectly rationalised in the current paradigm of scientific knowledge known as superstring theory. I shall begin this exposition with a look at the origins of the universe as we explore the multidimensional origin and nature of hyper-spatial eather and then I shall splice the fundamental essences of hyperspatial theory with superstring theory which will hopefully carve out the features of a new paradigm in their explication.

It is believed by physicists the world over that the big bang was the definitive starting point of the universe as it exists today; exists today as the fundamental matter from which we and others are made but there is more to the dimensional constitution of reality then would first meet the eye. Superstring theory has it that the fundamental dimensional nature of the universe is such that the universe exists in eleven dimensions all of which have causality in the fundamental constitution of the nature of the universe: especially in the determination of the fundamental constitution of physical matter as the vibrating circular string.

10

Paradigm Shift: Science & Technology of the Future

As such at the point in time when the proto-universe became unstable and started to unravel and drift apart at great speed, we see the four dimensions of space-time expanding out into the great cosmic abyss which we can only assume is infinite. I however believe that inter-galactic acceleration is caused by the missing 'dark matter' which exists at the edge of time and space in a symbiotic causal dynamic with the meta-spatial boundary which separates the inner existence of ourselves in our own eleven dimensions from whatever creature, form of life or kind of intelligence that exists in the twenty six dimensional universe outside our own about which ultimately we can say nothing. With the expansion of space-time came the contraction and collapse of the seven other dimensions; in this collapse and contraction of these seven dimensions into a layered mesh of doughnuts which penetrates and permeates all matter creating a hyperspatial tension local to each of the doughnuts. As such the doughnuts are circular in shape with a hollow centre; it is within this hollow centre where resides the hyperspatial tension in a state of balanced equilibrium vibrating a dynamic which consists of the frequencies which are essential to the properties and causalities of the hyperspatial eather in its fundamental determination as a part of both the

11

theory of the culture of scientific knowledge but also as a contributing factor existent in the fundamental causal dynamics which determine the nature of the world and the universe we inhabit.

Through the determination of the causalities of hyperspatial eather two distinct functions appear and are causal to our equating of or our conceptualisation of the fifth dimension and the causalities of eather as determined by the superstring theory. The first instance of hyperspace theory is that of the billion or so donuts contained within a small area of time and space exist as to be the causal determinant of the structure and shape of atoms. On the other hand the tension of this quazi-spatial dynamic is the root product of superstrings as they are constituted in their arrangement creating a medium through which light can travel both as a particle and as a wave.

The vibration which is made in hyperspace as such harmonises a pattern of the eather inside the electron shell which determines the properties of the electron shell/cloud constituents such as neutrons in reaction time dynamics and in superfluids. The resonance shell of the electron in its manifest dynamic exists from the causality of the inner skin which affects the causality of the electron shell through the vibration of the inner nucleus being a quantum determinant of the properties and causalities of the nucleus influencing the electron shells causal dynamics through the vibration of both protons and neutrons within the nucleus of the atom.

The first piece of technology which should be discussed first is the crafts propulsion device which enables the craft to make such moves as to turn ninety degree turns at speeds in excess of mach 6. If two electromagnets at opposite ends are turned on then they will attract each other. If one is stronger than the other then the energy dynamic results in a pulling force in the direction of the strongest magnet. When the two magnets are fixed in place then the force of the energy dynamic of the two magnets is transferred from the magnets to the chassis of the craft producing movement in a given direction. In order to change the direction of the craft is to change the direction of the magnetic alignment, to change the speed is to increase or decrease the amount of electricity running through the two magnets.

This explains how the craft moves but what is unknown still is how the crafts crew remain intact in these extreme changes of velocity and direction. What this piece of technology is called is the inertial displacement field. Inertia is the latent resistance of an object (inside or outside of gravity) to movement in a particular direction. Inertia is caused by the movement of an object through hyper-spatial eather in which the constituent substance of the eather is so small in size that they just pass right through us and penetrates all matter in the nature of a all pervading matrix of zero-point doughnuts which determine a hyper-spatial causality through the ramification of this zero-point matrix. The ramification is the hyper-spatial tension which is caused through the collapse and contraction of the six other dimensions into these doughnuts at the time of the expansion of the remaining dimensions of time and space at the point of the big bang. As such the higher dimensions still have their causality upon the nature of the manifest existence and essence in the manifestation from essence into existence of the four dimensions which expanded.

The existence of the universe is the unification of laws and forces which have come to exist and as such define both the logical and the causal five dimensional nature of the universe in its birth and in its continued existence. In the proto-universe there was nothing and given an infinite timeline the universes pre-universe unification of cosmic variables (such as the speed of light) came to produce a universes similar to our own but would eventually produce the existence of one dimension which is the perfect balance of the cosmic variables which governs the laws of nature and their manifestation and as such the totality of space still exists and is stable as the universe as we know it.

As such Fifth dimension is the tension of hyperspace through which the contracted six dimensions have their causality within the manifestation of the universe through the nature of the hyper-spatial tension. The all pervading tension of hyperspace is innate to the manifestation of any point particle (such as an electron or proton) from its inner state as a closed ended three dimensional vibrating string to its outer manifestation as a spherical atom with electromagnetic properties because of the nature of the electron shell. The point particles manifestation as smaller sub-atomic particles in atom smashers the world over produces a different number and type of particles varying with the speed of the two parent particles as they collide. This leads to the assumption in quantum mechanics that the point particle as it exists as the product of the parent atom as it is broken into different frequencies of the original cord of the atomic particle and as such broken by collision into different subatomic frequencies and therefore different point particles each and every time a collision occurs between two protons or electrons in an atom smasher. The skin of the particle be it atomic or subatomic is the product of the hyper-spatial tension being formed around the vibrating string of the particle manifesting as a taut skin of the surface of the atom or point particle. As such it is also true that certain wavelengths namely the wavelengths of light through the shape of the waveform being circular form a photon with the ultra violet and infra red manifesting photons the wavelengths of the electromagnetism being too close together or far apart for photon manifestation. This explains how plant life uses photons as electrons in photosynthesis and how certain vibrations be they of the strings of point particles or wavelengths of electromagnetic radiation either can produce a manifestation of matter from an hyper-spatial algorithm which determines the nature of the manifestation of matter from its base state as energy. As such hyperspace vibrates and it is on that premise I shall base my final technological argument that hyperspace can be both bent and folded and through the oscillations of an a/c current at the right wavelength and frequency through the displacement of the hyperspace in front of the craft creating a slipstream which the vessel can navigate. As such the displacement of hyperspace results in inertialess travel with the cause of inertia (hyperspace) being displaced around the craft.

All that is left to say now at the end of this discussion is that if eather can be warped around an object through the oscillations of an a/c current then there are two further ideas which can be assumed of hyper-spatial technology which is the radiation shield and the invisibility device. The radiation shield and the invisibility field are both ramifications of the same fundamental principle with is that with an a/c current oscillating through wires just underneath the armour of the vehicle or vessel can displace hyperspace with the vibrations unable to occupy the same time or space are bent around the vessel. Still the electromagnetic radiation though still follows the eather around the ship as though a bend in hyperspace had not occurred thus explaining the invisibility cloak, radiation shield and momentum and inertialess travel.

The inertialess field comprises of an electromagnetic fluctuation vibrating the same frequency as hyperspace which in a similar way to the ionosphere deflecting certain wavelengths of electromagnetic radiation and as such the electromagnetic field at certain wavelengths displaces the eather around itself as one cannot occupy the same space of influence of the other. This makes for a cloaking device producing invisibility in the object and a radiation shield as all wavelengths of electromagnetic radiation are displaced according to the displacement of hyperspatial eather.

Another of my ideas in this new paradigm is the weather control device which can cause rain in arid and dessert environments. There has been much contested doubt concerning weather devices and the effect they would have over the greater global climate and so it stands that an explication in the objective and subjective causal determinants of the essential nature of the global climate is necessary in order for us to advance. In the constitution of the global climate we come to see the weather as composed of weather systems which follows the path dictated by the movement of its originator: the ocean currents. As such the objective causality is the ocean currents and the subjective constituent is the weather we experience. A prime example of this is the Gulf Stream which makes our summers cooler and our winters warmer; the effect of this is profound as London is at the same latitude as Moscow yet experiences warmer weather. As the climate is the subjective consideration of the weather we can safely believe that little harm would be done to the ecosystem through the use of weather control

18

devices: as such what follows is a description of the fundamental mechanics and causal structures at work in a weather control device.

Two things exist in dessert and arid climates these be static and moisture (due to the intense heat the water exists in a form of suspension) It is the static which is the primary causality here as all static vibrates at a certain frequency. If you place skyward pointing magnets in the arid environment each having a zone of effect which overlaps with the other magnets creates an area in which the moisture may be brought out of suspension and form rain clouds.

How this is done is through the magnets having a oscillating AC current running through the wires which supply the core of the magnet its electromagnetism. As the polarity of the magnet shifts from north to south the flipping of the poles of the magnets creates in effect a waveform which is projected into the atmosphere. When the waveform hits the correct frequency (the frequency of atmospheric static) then a coherent body is formed as the static not only vibrates with the magnet but vibrates with itself creating a coherent body of causality. As such this matrix or net of static is exactly what is needed to bring the latent water in the atmosphere out of its high energy suspension and thus form in its low energy state form clouds and rain as the electrostatic force generated by the magnets acts as a buffer absorbing the energy of the water particles.

The worst nightmare of any physicist is the black hole, a planet threatening asteroid with the right thinking can be easily obliterated but a black hole as an indestructible threat is off the scale as an extremity, even giving all of the worlds current technology the forces required to destroy a black hole are so extreme that there is no hope for survival that is before the nuclear fusion bomb. The nuclear fusion bombs delivery system is the starting point of this discussion where I explain the ways in which a black hole if discovered can be destroyed through technology not that much more advanced than the present day can be employed to save humanity from the forces of nature that would be the black hole.

The delivery system of the fusion bomb is what I call the magnetic acceleration tube. This is where sequenced magnets in a line fire when the fusion bomb is in the convergence arc of the magnet pulling the fusion bomb along at progressively faster speeds through the triggering of lasers until the speeds approach the speed of light. The first thing to note is that in space the coldness reduces the resistance of the magnet making the magnetic field stronger as the current can move faster through the wire, what should also be noted is that because of the temperature of space acting as a coolant allows much higher voltages producing an even stronger magnetic field. The final thing to note about the delivery system is that because of its construction in orbit the gravity is zero therefore the pull of the magnets has an astronomical effect on the potential speeds of the fusion bomb because the bomb or otherwise object being propelled has no weight, there is no resistance to the effect of the magnet thus allowing greater speeds at a shorter length of the sequenced magnet tube.

For a fusion bomb what you need first of all is atomic hydrogen which can be obtained through a chemical reaction which is released into a vacuum. What happens next is that the atomic positively charged hydrogen is released into a vacuum which collects the hydrogen and stores it within magnetic fields generated outside of the vacuum which holds the hydrogen for a brief while whilst it is implanted within a hollowed out diamond which by laser convergence is sealed shut. As such in order to initiate fusion a group of lasers all firing from different angles converges on the hydrogen itself whilst two magnets with positive ends facing the diamond spin round in opposite directions hoping to create an atomic friction out of the positively charged hydrogen atoms generating enough heat and pressure to cause hydrogen fusion to take place. This however is not the true scale of the weapon; the excess hydrogen is fused under the heat and pressure of the first hydrogen fusion and all the hydrogen fuses into helium creating a truly apocalyptically sized explosion which would be on the same scale as the size of the sun. Such a weapon if constructed on planet earth would explode with such force as to reduce the planet to an asteroid belt of rubble but for the purposes of destroying a black hole it would be entirely effective with the colossal forces required being demonstrated by the cluster fusion of the device.

The Podkletnov gravity shield works in a very simple way a super-conductive disk is spun at 6000 rpm and reduces the gravity effect of all objects above the gravity shield. At even faster speeds such as 13000 rpm the disk takes off, the question then is how we theorize this behaviour with our understanding of how gravity works. This picture of how gravity works begins first with how our theory of the cosmos is and how the universes expansion is not just expanding but actually speeding up. This acceleration of the cosmos is a somewhat unaccounted for element of our cosmological picture but it is my aim to reconcile this with a functional new theory of how gravity works as such which accounts for the functioning of the disk within a greater cosmological picture.

Due to the speeding up of the cosmos's expansion it can be postulated that there is some mass or somebody at the edge of space and time that we are accelerating towards. If there is some mass at the edge of time and space then this would account for the universes acceleration without having to appeal to the creation of theoretical abstract forces such as dark force, the counterpart of dark matter currently within the modern paradigm of physics thought to be responsible for the acceleration of the cosmos.

Gravity is the product of two forces the graviton within the atom which is also known as the gluon matrix and a higher dimensional point particle called the gravatino named as such because its speeds are vastly in excess of the speed of light as it moves through a higher dimensional causality. The gravatino is unique to the atom; there is one gravatino for each and every atom which constitutes all of the matter in space-time. Because the gravatino moves as a point particle in a higher dimensional space its movement in the lower dimensions which we inhabit is as a concentric movement from the edge of time and space where all is connected in unity through all of matter within the sphere until it reaches the gluon matrix within the atom. As the graviton moves through matter it picks up gravitational energy, this energy is released as a spasming of the gluon matrix towards the object of nearest gravitational pull. As the gravatino moves through the higher dimensional structure of the superstring donuts of the collapsed dimensions it creates a vibrational charge and frequency, this frequency can be tapped through harmonization which like a guitar string absorbing energy through the strings when a violin is played nearby allows for the transfer of energy through vibration. The movement of the gravatino through the higher dimensional space also accounts for the weakness of the force as it has all that work to do with gravatino charge slowly dissipating the further you are from the source of gravatonic charge.

The Podkletnov disk is basically a superconducting disk which when revolved at high speeds creates the effect of a gravity shield for all objects above the disk. This is because as the disk is spun around the lack of an electric field or electron vibration means that the more subtle vibrations of the atoms when revolved can accumulate eventually reaching a resonance which the gravatonic charge moving through hyperspace harmonizes with: the spinning disk leading to the absorption of gravatonic energy by the disk through the energy transfer of harmonization. Once the harmonization is achieved the gravity is absorbed by the disk and at a point of saturation the disk takes off in a funnel of gravatonic energy created which spins above the disk and acts like a cyclone. Saturation and take off are only achieved at the higher speeds due to the vibration of the entire disk with residual gravity as opposed to just the disks edges. This explains how the disk reduces the weight or mass of everything above it and explains how the gravity shield works without resorting to infinities for the explanation of the point particle explanation of gravity: the gravatino. The question is: is it the mass of the object above the gravity shield which is reduced or is it the weight of the object which is reduced. If it is the weight which is reduced then the story ends there but if it is the mass which is reduced then the gravity shield has its uses as a part of the fusion bomb package. As the fusion bomb descends into a black hole there is a spacetime distortion caused by the intense gravity field, the question is with enough Podkletnov disks could there be a cumulative effect where all the gravity is absorbed and the fusion bomb be unaffected by the gravity field of the black hole.

France in the last twenty years has commenced with some subterranean nuclear weapons testing on coral atolls in the Pacific Ocean. When one of these fission bombs is detonated it leaves a large hollow vesicle surrounded by tough solid glass made from the heated and displaced bedrock in the explosion. This glass is at least a Kilometre thick and is sturdy enough to take the weight of the coral atoll above the vesicle and support its weight for a long period of time without cracks in the glass appearing or any form of structural compromiseation. These huge underground vesicles could be used as an international solution to global waste management all that we need to do is drill our way through the thick glass near the surface to reach the inside of the vesicle where there is a huge space for the dumping of waste both toxic and non-toxic.

Circular momentum in an electromagnetic field excites zero-point ether in such a way so that virtual electrons embedded in hyperspace manifest into true electrons from the electromagnetic field leaving the wire in the electromagnetic field as it rotates or as the wire in the magnetic field as the magnet spins to excite through oscillation through the spin of the magnet up these newly quazi-manifested electrons. This is possible through only certain metallic elements and represents the spectral lines of the elements electron shell as a absorber and emitter of electromagnetic energy. The idea now is to create a spinning magnet similar to a electromagnetic generator but which loses none of its energy as it spins creating a device which approximates perpetual motion.

To begin with you need a sealed vacuum for a bar magnet to rotate within. If a hole is drilled in the centre of a powerful bar magnet and another bar magnet is threaded through then at either end two magnets of corresponding poles can be used to levitate the spinning magnet leading to no loss of energy from the of the magnet as it spins within a vacuum; leading to no loss of energy through air resistance. As such as long as the transfer of energy from the spinning bar magnet to the wire in the electromagnetic field as the magnet spins does not deplete the energy of the spinning bar magnet then something similar to perpetual motion is possible. None of its energy in the transfer of energy from magnetic spin to the wire is lost, if this is so then the bar magnet should lose none of its energy leading to it spinning indefinitely, at least in theory. Due to the sealed environment of the spinning magnet (the zero-point oscillator) can be created using a series of magnets which attract the poles of the oscillator on the outside of the sealed environment creating the spin of the primary bar magnet. This spinning magnet is initially used to power a battery which then fires electromagnets which increase the spin of the hyperspatial oscillator (the central spinning magnet) if it should ever decrease in spin through an unforeseen factor that constitutes a loss of energy with the magnet spinning in a vacuum suspended in mid air by magnets causing the floatation on cushions of electromagnetism. If the theoretical physics behind this vacuum generator is correct then this device is a perfectly viable way of tapping the electromagnetic ether for energy as the vacuum generators energy output as electricity is greater than the input of rotational spin; energy is neither created or destroyed: the zero-point hyperspatial eather is tapped for energy and this factor constitutes the greater output then the input of the essential causalities of the theoretical schematisation of the technological device. The energy dynamics of the device are as such: as the magnet gets spinning the energy of the magnet spinning is stored in batteries which charge the magnets which are used for the spinning magnet so the machine never needs recharging. Also as a further idea is that as the speed of the magnet increases so too is the suspension magnets either side of the zero-point oscillators (the magnet in the middle which spins) energy output increases allowing progressively faster speeds as fast as the electricity of the acceleration magnets can flow from the laser detector of the magnets presence to the next magnet which is responsible for the zero-point oscillators accelerator to truly atrocious speeds, at least in the theory this is so.

Solid fuel is a means of rocket propulsion which has yet to come fully into effect in the technology of rockets, the science is very simple: as you descend the periodic table of elements the denser the atoms are the more violent the reaction which leads us to believe that if we consider solid fuels for rockets then we have a working template to the development of these fuels being that if we make solid fuels the further down the periodic table we consider the constituent molecular element; the more vigorous and energetic the energy release: The molecular structures as such are possible through the causalities of chemical science. The next device I shall discuss is the hyperspatial drive and how it could be used in faster than light travel.

The nature of the hyperspatial eather is such that there are two different manifestations of it: One trans-dimensional which exists outside of conventional space and time which I call solid eather and one form of eather which is quazi-dimensional in nature which exists partially inside and partially outside our spacetime lower dimensional reality but as such is causally linked to the trans-dimensional the quasi-dimensional being the ramification of the trans-dimensional in the constitution of the dimensional factors which as such govern the nature theoretically linked causally connected physical universe. It is liquid eather that allows for the movement of light through space and the atmosphere through a property of the hyperspatial eather called locality. Non-superstring materialised eather is localised to a particular donut of 'solid' hyperspace which if it becomes displaced the liquid quanta desires to be at its point of origin; this creates the elasticity necessary to transmit light, heat and all other kinds of electromagnetic particles and vibrations as they are transmitted as all quasi-dimensional eather is harmonised with itself creating a coherent solid body through which light can travel.

An atom as such is the physical manifestation of a higher dimensional closed ended string which vibrates at a certain frequency. Quasi-dimensional hyperspace vibrates at a number of frequencies and where there is a causal resonance between the string and hyperspace a skin is formed around the vibrating cores frequency which through the structure formed by the hyperspatial dynamic becomes the vibration of the electron shell/cloud and hence a skin is formed around the vibrating string through a multidimensional harmonic of eather with superstring thus explaining and demonstrating the connection and the causality which binds the two theories together.

When the vibration of a photonic electromagnetic wave travels through the eather and at a certain point the wave creates a circular shape and through the three dimensional wave form as it is manifests because of its frequency where a spheroid manifests due to the nature of the eather forming a skin over the hyperspatial vibration as it travels from origin to destination. The next target of investigation in this elaborative exhibition of what new science is and can be capable of is the faster then light hyperspatial drive.

The structural design of the craft for hyperspatial travel is either a disk or a teardrop, I shall explain why. Using magnets and wiring an alternating current a vibration can be found which is at the exact frequency of the hyperspatial eather. The a/c magnets oscillation and hyperspace cannot occupy the same space and time just like long wave radio waves which are reflected off the ionosphere the two are immensurable.

As such the displacement of eather around the craft as it moves creates with the liquid eather a pressure around the craft as the liquid eather through locality creates a pressure around the sides of the craft but due to the presence of more electromagnets the displacement through the craft to their localised origin cannot occur and instead creates pressure on the surface of the sides of the vessel through pressing against the electromagnets. Due to the shape of the tail three hundred and sixty degrees around the edge of the disk and the edge of the tail of the teardrop the hyperspatial tension forcing against the slope creates motion. Thus the hyperspatial drive utilises hyperspatial tension caused by the displacement of eather to affect the motion of the craft.

The next area of is discussion is artificial blood. Blood banks and donations exist because we have not yet realised how we can make it synthetically, this along with so many futuristic new technologies need the use of the precision that robots have potentially in their capabilities and capacities in order for the item or object to be made or produced. Ultimately it is the limits of detection, programming and action all tied into each other in a complex way.

As such what I call super science relies fundamentally upon the mode of production in the orientation of the goals and agendas of the states bourgeoisie (Karl Marx) and the limits upon the world through the worship of illusory objects or gods in false consciousness takes our sentiment from the gross inequality and oppression in the world. The North Atlantic Treaty Organisation or NATO or the United Nations the UN must act as to protect the sacredness of liberal freedom, of life and the equality of all human beings.

So it is then that the mode of production satisfies only the investors and only they have any real power. The question is however unlike the proletarians who forged socialism in the communist revolutions do we have anything to lose such as our civil liberties and human rights?

The age of super-science as I call it is the stage of technological progression, the detail and sophistication of scientific theory and a deep emersion of the human life within the technological ramifications and inventions of the present day.

For the next stage of super scientific refinement and plateau of emancipation and freedom to flourish which is in the next 'paradigm shift' (T. Kuhn) As a pinnacle act of the foundations of neo-modernism the new possibilities created through technological advancement will usher in liberation from the forces of nature, create liberal freedoms and restructure the international political autonomy and economy so that the stock market is governed by supply and demand: not profit and exploitation.

The final idea for this book is a new design of body armour: based on mathematical principles. Everybody knows if you push the top and bottom of an egg together then the egg will not crack or break. How you produce the eggs from steel of such a small size is through allowing tiny droplets of steel dripping into water that causes solidification; at a certain height the droplets form into egg shaped iron droplets formed when the liquid iron in falling tries to regain its equilibrium against the force of the air resistance as it falls.

These eggs are mixed in to molten Kevlar which has a lower melting point then the iron so the eggs stay in shape. What happens next is that the liquid Kevlar is pressed into thin strips then at a precise viscosity the Kevlar passes under a very long magnet which as it cools aligns all the steel eggs into formation ready to absorb the impact.

www.ingramcontent.com/pod-product-compliance
Lightning Source LLC
Chambersburg PA
CBHW021854170526
45157CB00006B/2437